Endeavour Views the Earth

NASA's Lyndon B. Johnson Space Center (JSC) is the home of the U.S. Space Shuttle program. At JSC, the Space Shuttle Earth Observations Project within the Flight Science Support Office trains NASA astronauts in Earth observation techniques, and catalogs and analyzes the astronauts' photographs after each Shuttle mission. JSC's Imaging Sciences Division processes the film from the Space Shuttle missions, and distributes copies to photo data centers throughout the country. The Training Division in JSC's Mission Operations Directorate provides camera technique training to the astronauts.

Endeavour Views the Earth

Compiled by
Robert A. Brown
Space Telescope Science Institute
with the assistance of the *Endeavour* Astronauts
NASA Johnson Space Center

CAMBRIDGE
UNIVERSITY PRESS

Published by the Press Syndicate of the University of Cambridge
The Pitt Building, Trumpington Street, Cambridge CB2 1RP
40 West 20th Street, New York, NY 10011-4211, USA
10 Stamford Road, Oakleigh, Melbourne 3166, Australia

First published 1996

A catalogue record for this book is available from the British Library

Library of Congress cataloguing in publication data available

This book is a publication of the Exploration in Education (ExInEd) program of the Special Studies Office at the Space Telescope Science Institute. ExInEd's purpose is to increase the educational value of America's exploration of space by finding new ways to communicate its attendant ideas and results. A version of this book is also available in electronic format. For more information, please write to: ExInEd, Space Telescope Science Institute, 3700 San Martin Drive, Baltimore, MD 21218. (ExInEd@stsci.edu) Support for this research was provided by NASA under Grant NAGW-3048 to the Space Telescope Science Institute, which is operated by the Association of Universities for Research in Astronomy.

Design by Foxglove Communications

Printed in the United States of America
Contents used with permission

ISBN 0 521 57099 9 paperback

Contents

Introduction

When astronauts first went into space, the government had no plans for us to take cameras. The pilots had other ideas—they knew that one way to capture a fraction of the glorious moments that occur when you rise above the Earth is with a camera. One of the first seven astronauts (Wally Schirra) was an accomplished amateur photographer, and he personally bought the first camera carried into orbit by an American. Official acceptance was quick, because the photos were spectacular!

In our opinion, there is no better way to share the experience of spaceflight than by poring over photographs. The U.S. Space Shuttle has, by far, the best expanse of windows of any spacecraft yet built. In fact, the view from inside is not very different from the view one has being outside on a spacewalk. The photos we take are, for the most part, natural color (we do use a small amount of infrared film for special purposes). What you will see in these pictures is what you would have seen out the windows had you traveled with us.

The Crew of STS-47

Seven of us climbed aboard Endeavour on September 12, 1992. Three of us had responsibility for the Orbiter: Robert L. "Hoot" Gibson, the commander; Curt Brown, the pilot; and Jay Apt, the flight engineer. Our cargo—a large laboratory for conducting physical and life sciences experiments—was crewed by Mark Lee, payload commander; Jan Davis, mission specialist; Mamoru Mohri, payload specialist; and Mae Jemison, science mission specialist. This photograph shows all of us as we left the crew quarters at Kennedy Space Center in Florida to head for the launch pad.

The Launch of STS-47

Endeavour launched right on time into an orbit that carried us as far north and south as 57 degrees latitude. Our altitude above the Earth was 187 miles (305 km). We launched in mid-morning, which meant that we flew over the Northern Hemisphere in sunlight and the Southern in darkness throughout our mission. As a consequence, you won't find any photos of South America or Australia in this collection.

Cameras on the Flight Deck

This is the aft portion of our flight deck, and shows the three 70 mm format Hasselblad cameras that we used, as well as two of the ten windows. In addition to the Hasselblads, we had one 4 inch by 5 inch format camera and a 35 mm camera with a 300 mm telephoto lens for special shots. We operated around the clock in two shifts, with at least one of us on the flight deck except for brief periods.

So, what is it like up there? Well, weightlessness is lots of fun, but what really grabs your interest is the Earth out the window. Most of us spent every free moment looking outside. We took over 5000 photographs of the Earth during our 8-day flight. That averages out to about one every 30 seconds for the time we were over land!

We have chosen less than 1% of our pictures to show you. The dominant criterion used in selecting them is whether they will help you share the experiences of spaceflight. We say "experiences" because there are at least three separate ones that we would like to share with you.

First, a sense of geography. Viewing Earth from the Shuttle allows you to see the planet almost as a living map. You can take in all of New England at once, and discern for yourself why the irrigation patterns in the Saudi Arabian desert must be following underground rivers. It is amazing to see how small the Middle East and Europe really are.

Second, a sense of what forces are at work on the Earth. Hurricanes, ocean currents, aurorae, and dust storms are apparent. The forces that slowly raise the land and drain seas are all easy to visualize up there, and by merely looking out a window, we can see such sights as the crumpled "fender" of Pakistan's Makran mountains caused by India's geologically recent smash into Eurasia.

Third, a sense of the changes nature and humans cause on our planet. Both natural pollution from volcanos and human changes are evident. Poor decisions by people have had consequences that have affected what we see from orbit, such as the dramatic shrinking of the Aral Sea. But the dominant impression is one of great beauty, and most of us return from spaceflights with a feeling that Earth is a holy place, unlike any other planet in the known universe.

The pictures in this book are organized geographically, going from west to east (the way we fly around the Earth). The final four photographs show the aurora and atmosphere of our beautiful planet.

Enjoy your flight!

Mount St. Helens and Mount Adams, Washington

46.2 N 122.2 W

We start our tour on the West Coast of the USA, with a dramatic indication that the Earth is, geologically, a very active planet indeed. Enormous forces are at work under the thin crust of the Pacific Northwest.

This low-oblique view shows two of the major volcanos of the Cascades of southern Washington: Mount Adams at the top right, and Mount St. Helens at the center left of the view. Mount St. Helens erupted on May 18, 1980, removing 1300 feet from the summit of the 9677-foot volcano. The eruption toppled trees with a searing, stone-filled 275 mph wind over an area of more than 150 square miles. This area, now called the blast zone, can easily be seen in the view. Natural regrowth of vegetation within the blast zone is progressing at a rapid rate, especially on the outer fringes and in the protected valleys. Many fir trees have grown to heights exceeding 20 feet in a little over 12 years. Clear-cuts, a major environmental concern to many people in the Pacific Northwest because of problems of erosion and lost habitats, can be seen speckling the forest lands in the view. (STS047-96-065)*

* This number is the order number for the photograph opposite.
  (See page 82 for ordering Space Shuttle photographs.)

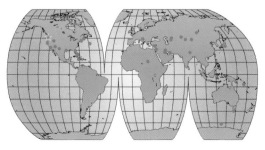

Mount St. Helens, Washington

46.2 N 122.2 W

This photo was taken as we flew directly over the top of the mountain. The picture covers 41 miles (67 km) a side. The tail of the Space Shuttle Orbiter cuts across the top of the picture. A large lava dome within the crater of the volcano has grown to a height of over 1000 feet since the 1980 eruption. All of the major volcanos of this region—Rainier, Adams, St. Helens, Hood, and Jefferson—were created by explosive volcanism, whose recent results are seen in the picture. (STS047-96-069)

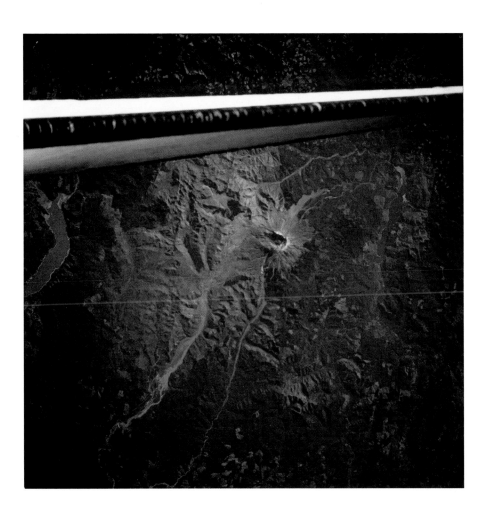

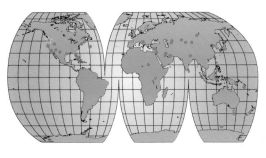

Entire California Central Valley

38 N 121 W

The fertile central valley of California, lying between the coast range and the high Sierra, provides a large proportion of the vegetables eaten in the United States. We took this photo from near the California–Oregon border, looking south. Notice the absence of snow in the Sierra, due to the severe drought affecting California at the time of our flight. The basin-and-range territory seen to the east of the Sierra is due to the cracking and dropping of blocks of crust as the countryside was lifted upwards in the collision of crustal plates. (STS047-153-358)

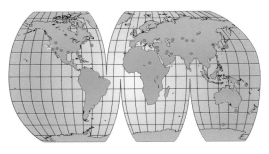

San Francisco Bay and Monterey Bay, California

37 N 122 W

A California drought has turned the hills of the Coastal Range orange in this photo. Only the Santa Cruz and Santa Lucia ranges along the coast were green in September 1992 when we flew over the area. The light gray around the San Francisco Bay is the tell-tale signature of cities. Salt ponds can be seen at the south end of San Francisco Bay, and a turbid mud flow is visible in the bay. Monterey Bay is slightly below center, with Santa Cruz at the north end of the bay, and Monterey at the south end. (STS047-151-083)

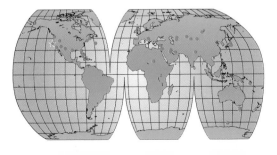

Mono Lake, California

38 N 119 W

This lake just east of Yosemite National Park has been drying up for thousands of years, and some of the ancient shorelines can be seen in this photo. Recently, the diversion of California rivers to agricultural irrigation in the central valley and the population centers has accelerated this process, exposing limestone pillars 20 feet high and the white salt of the lake bottom. Due to a prolonged drought in California, there is no snow cover visible in the high Sierra. This picture is 41 miles (67 km) a side. (STS047-94-032)

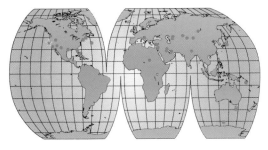

Central Valley Irrigation System, California

40 N 122 W

The dark mountains of the coast range and the Sierra border the fertile crop-growing central valley of California. Water has always been the key to the economy of the western United States, and this sun-glint photo shows a portion of the extensive irrigation system that draws water from all over the West for California's farmers. Irrigation has allowed excellent crop yields, but the minerals in the water are gradually accumulating in the soil. Recently farmers have had to flush the fields with water to wash away salt buildup, and more frequent flushing may be required in the future. This picture was taken looking to the west. (STS047-75-007)

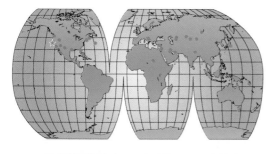

Los Angeles

33.5 N 117.5 W

Cities are gray as seen from orbit. Los Angeles and its suburbs extend nearly the full 104 miles (167 km) covered by this picture. Point Mugu is in the upper left, with the dark Santa Monica mountains to its east. The eastern end of the mountains is the Hollywood Hills, and the 101 freeway can be seen running from Los Angeles to the San Fernando Valley. In the top center of the frame are the San Gabriel mountains (overlain by an orange window reflection—we were looking into the Sun when we took this picture). This was a clear day; often the mountains hold in dense smog. Strict air pollution controls have increased the number of clear days in the last decade. The 4000 foot high Edwards plateau lies to the north, in the upper right. Palos Verdes provides the one spot of green on the coast, pointing toward Santa Catalina and San Clemente islands. Inland, the dark Santa Ana mountains are bordered to the east by one of the numerous faults along which the Earth's crust continues to move. (STS047-84-074)

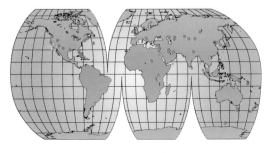

The Grand Canyon, Arizona

36 N 112 W

As the western part of the United States gradually pushed over the top of the Pacific continental plate, Arizona was lifted thousands of feet above sea level. During this time, the Colorado river largely kept to its course, and cut this fantastic canyon through the uplifting Earth. The steep walls of the inner gorge, cut through hard granite, are visible along the river's course. Softer rock above has crumbled into the canyon, leaving wonderful shapes. The tail of the Space Shuttle points to the Bright Angel fault, a crack running across the canyon, which formed as the land rose. The south side of this fault provides one trail into the canyon. Dark national forests show up on both the south and the wetter north side of the canyon. This photo is 104 miles (167 km) a side. (STS047-99-042)

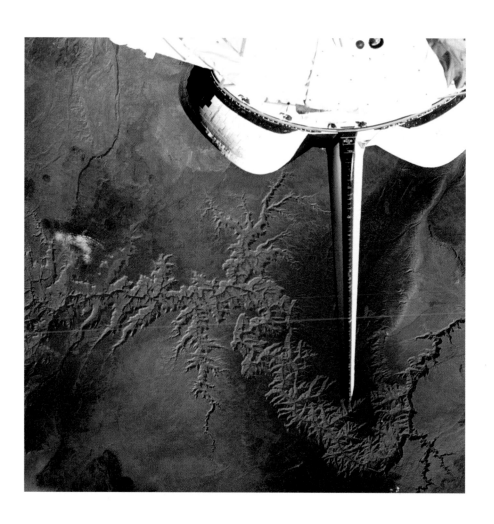

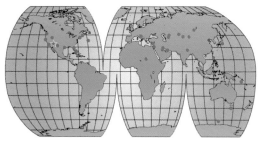

Great Salt Lake, Utah

41 N 112.7 W

What a sight from orbit! This is the remnant of a huge sea that once covered much of the United States' high plains. As the western part of the continent collided with the Pacific plate, the crust was buckled and uplifted, and most of the sea drained into the Gulf of Mexico. What's left is the Bonneville Salt Flats, to the left in this photo, and the mineral-rich Great Salt Lake. Salt Lake City is to the lower right of the lake, and Ogden is at the center right, near the bright evaporating ponds. Railroad trestles divide the lake into three parts, each with a different salinity (and color). This photo is 104 miles (167 km) a side. (STS047-97-021)

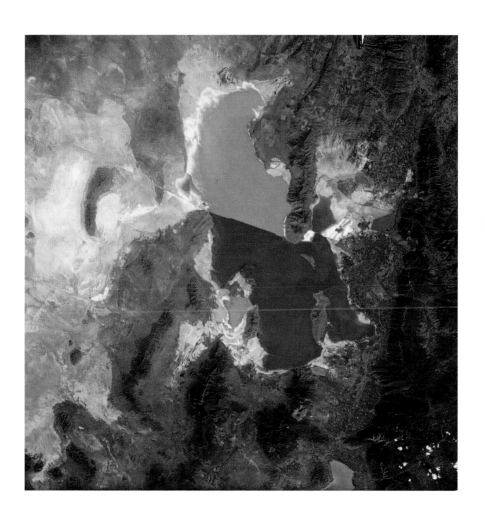

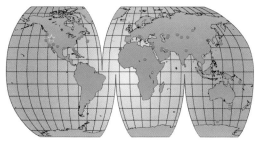

White Sands, New Mexico

32.8 N 106.2 W

This gypsum field 75 miles (120 km) north of El Paso, Texas, is the remnant of an ancient lake, which dried up as the land was lifted up by the collision of tectonic plates. It is a prominent feature, even on pictures taken by the Apollo astronauts on their way to the Moon. The 8000 foot high San Andres mountains are to the west (left), and Holloman Air Force Base and the town of Alamogordo can be seen on the right in the foothills of the Sierra Blanca. This hard gypsum provided a landing field for Columbia on the third Space Shuttle mission, when bad weather closed the dry lakebeds at Edwards Air Force Base in California. (STS047-76-055)

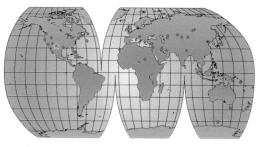

Galveston Bay, Texas

29.5 N 95 W

Houston is home for those of us who fly into space for the United States, and we tend to take lots of pictures of it. This is an unusual photo, as the glint of the Sun shows the smallest details of our watery environment. West is up in this picture, and North to the right. Starting at the bottom, we can see Sabine lake, where scores of offshore oil rigs are tied up awaiting work. A portion of the intracoastal waterway, built as a defense against Axis submarines in World War II, runs inside the barrier island to the large heart shape of Galveston Bay. You can see Galveston Island farther along the coast. It will come as no surprise to you that the entire island was put under water by the great 1900 hurricane. Galveston Bay is only 10 feet deep, so a channel is dredged for large ships. The chain of islands in the bay marks one edge of the channel. Ships can go all the way up the waterway shown in the upper right of the picture. This photo was taken in late afternoon in September, and the characteristic Houston thunderstorms have built to the west of the city. (STS047-74-101)

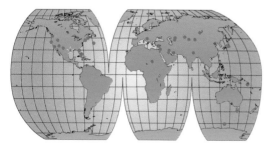

Sun-Glint on Southern New England

41.5 N 71.5 W

We took this photo when we were east of Cape Cod looking west into the Sun. Georges Bank, a shallow area of great commercial fishing value, is in the foreground. The sun-glint illuminates internal waves here just under the ocean surface. Cape Cod, Nantucket, and Martha's Vineyard are terminal moraines—literally piles of rock debris left at the southernmost edge of a great glacier that scoured New England when the Earth was much younger. Buzzard's Bay and Narragansett Bay are clearly outlined by the Sun. Long Island and the Hudson River are silhouetted in the center of the picture. (STS047-151-098)

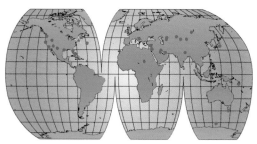

Kennedy Space Center, Florida

28.5 N 80.7 W

The first rockets were launched from "The Cape" (Cape Canaveral) in 1951. Its location has three advantages: (1) any malfunctions (and there were plenty in the early days) can be dealt with over open water, not endangering cities; (2) it is far enough south to take advantage of the 750 mph speed of the Earth's easterly rotation at this latitude (that is 3% of the speed needed to get into orbit, and it is free!); and (3) very large items can be brought in by boat. The actual cape is dotted with Air Force launch pads, while the NASA Space Shuttle facilities are farther north, just above the smaller bend in the shoreline. We landed Endeavour on the three mile-long runway visible in the photo. The tail of the Space Shuttle Orbiter is in the left half of the picture. One side of the photo covers 41 miles (67 km). (STS047-80-076)

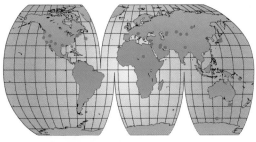

The Caribbean

23.5 N 76.5 W

After the exhilaration of flying over our own country, the beautiful Caribbean
provides a last treat as we head for the open Atlantic. Shallow limestone
(formed from coral) and clear water make it as easy to see the landscape under
the water as on the islands above it. The light blue carbonate banks of the
shallow Caribbean near Great Exuma island spill into the 3000 foot (1 km)
abyss called the "Tongue of the Ocean" in the upper left portion of this picture.
The shallow waters are rich in oxygen, so they have abundant life. The deep
waters of the abyss do not mix rapidly with the shallow waters, so life is
much rarer there. (STS047-153-038)

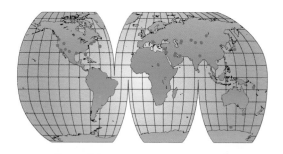

Western Wall of the Gulf Stream

43.6 N 65.4 W

On this mission, the orbit of the Space Shuttle took us over the eastern sea-board of the United States and out over the Atlantic Ocean just south of Nova Scotia. The orientation of the Shuttle afforded us an excellent view of the sun-glint pattern as it moved across the surface of the ocean, highlighting ocean phenomena that affect sea surface roughness. In this view, the western wall of the Gulf Stream is clearly visible, running diagonally across the photo. The sharp line defines the boundary between the warm, rapidly flowing waters of the Gulf Stream (right portion of the photo) and the cooler, coastal shelf waters (left portion of the photo). Notice the difference in the apparent dynamic processes in the two water masses: the smooth, nearly featureless, appearance of the shelf water compared with the complicated eddy field in the Gulf Stream water. (STS047-80-086)

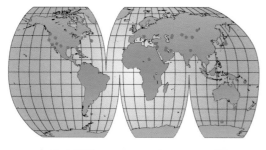

Hurricane Bonnie

35.4 N 56.8 W

On the last full day of the flight, September 19, 1992, four of us were on the flight deck when mission control called from Houston. They told us that in five minutes we'd be flying over a dwarf hurricane 500 miles northeast of Bermuda, and asked us to photograph it. We programmed our laptop computer with the coordinates, and we sighted the hurricane exactly when the computer said we would. You can see its eye and spiral bands clearly in this photo. Fortunately, it never hit land. (STS047-151-618)

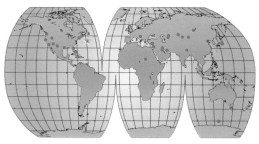

Italy and the Adriatic

44 N 15 E

We flew in mid-September 1992, when a strong temperature inversion settled in over Italy and the Northern Adriatic. This inversion acted as a cap to keep the industrial pollution from escaping upwards. The Alps to the north are high enough to keep the smog out of Northern Europe, but Italy was so smoggy that it was just barely visible even from directly overhead. The tops of the Apennine mountains down the long spine of Italy are marked by cumulus clouds. Milan is in the thickest part of the smog just south of the Alps, and Genoa is just south of the western edge of the Alps. (STS047-99-013)

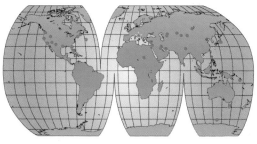

Southern Turkey and Cyprus

35 N 33 E

Old oceanic crust that has been heaved up to the surface is visible as the black rock bands on the island of Cyprus. The line separating the Greek and Turkish populations parallels the thick black band, and is slightly to its north. Turkey is the landmass under the Space Shuttle Orbiter's tail. The area covered by the picture is 200 by 250 miles (320 × 406 km). (STS047-153-271)

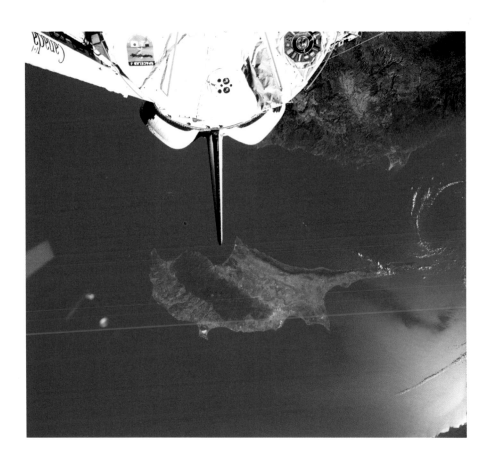

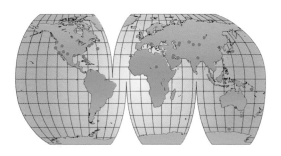

Mourdi Depression, Chad

18 N 22 E

Before we fly for the first time, most of us have an image of North Africa as filled with endless yellow sand dunes; the reality is fantastically varied oranges and even reds. Here in northern Chad, jet black wind-scoured volcanic rock lies beside brush strokes of burnt orange and cadmium yellow from an artist's oil palette. The rivers of sand are pushed by fierce desert winds, and almost seem to be flowing in this photo. The area shown here is known as the Mourdi Depression, near the border between Chad and Libya, and is one of the most memorable sights in the Sahara. (STS047-151-040)

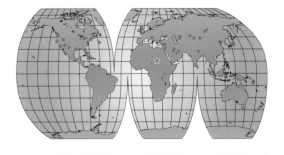

The Levant

31.5 N 35.5 E

Blowing dust obscured parts of the Middle East during most of our mission, but this day was relatively clear. The Red Sea branches into the Gulf of Suez to the right (west) and the Gulf of Aqaba to the left (east). The Dead Sea, Jordan River, and Sea of Galilee lie in the Dead Sea Fault, ending in the mountains of Lebanon. We can see borders from space only rarely, but the Israel-Egypt boundary in the Sinai shows up because the Israelis irrigate and grow crops on their side, and the Egyptians do not. South is up in this photo. (STS047-101-039)

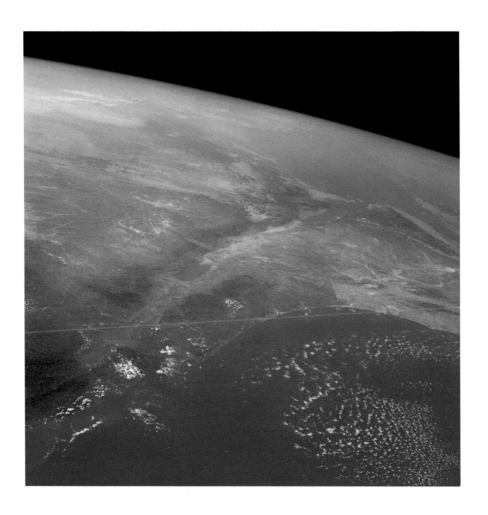

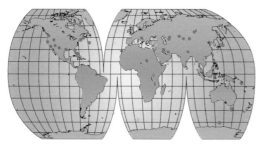

The Dead Sea, Amman, and Jerusalem

31.5 N 36 E

The Jordan River comes in from the north, through the Dead Sea fault. Amman, capital of Jordan, is the gray area just above and to the right of the center of the frame; its airport is about 15 miles south. Jerusalem is halfway from the west (left) edge of the Dead Sea to the edge of the frame. This picture covers 70 by 91 miles (115 × 146 km). The salt pans at the south (bottom) end of the Dead Sea were part of the Sea as recently as 1965, but irrigation has lowered the water level enough to expose the land bridge. (STS047-151-318)

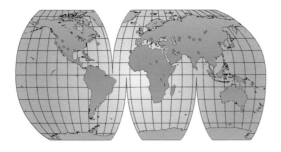

Oil Fire Residue in Kuwait

29.2 N 48 E

In early 1991, invading Iraqi troops set the oil wells of Kuwait ablaze during their retreat, sending up huge clouds of thick black smoke. Fortunately, the smoke was heavy enough that it never reached higher than 15,000 feet (5 km), staying below the strong winds that would have spread it beyond the Middle East. The fires were extinguished 8 months after the liberation of Kuwait, but the scars remain. The sand of the oil field in this picture was blackened by the soot and tar from the fires. This picture covers an area of 15 by 23 miles (24 × 36 km). (STS047-53-01)

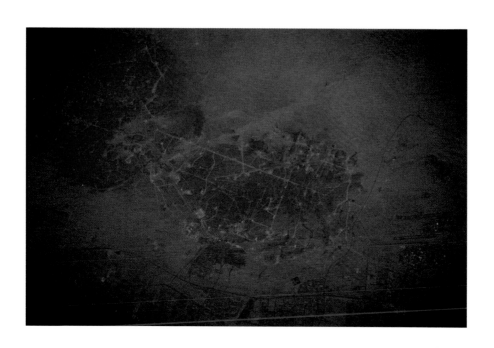

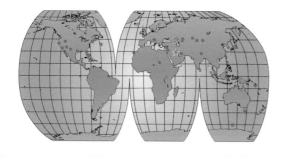

Center Pivot Irrigation in Saudi Arabia

26 N 47 E

This location is roughly 60 miles (100 km) southwest of Riyadh, Saudi Arabia. Each circle visible in this photo is a farm. Irrigation is provided by drilling a well in the center of each farm, and running irrigation lines out on moving supports, which sweep out the circle over time. The pattern of the center-pivot irrigation farms can be seen to follow underground rivers. Though the water in these rivers makes the desert bloom, it may last only 50 years. This picture is 41 miles (67 km) a side. (STS047-91-084)

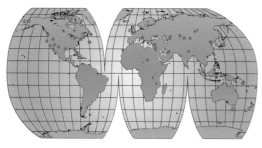

Zagros Mountains, Iran, and the Strait of Hormuz

28 N 54 E

The steel-gray curves of Iran's Zagros mountains are one of the most memorable sights for a space traveler. The range reaches 13,000 feet, and is a product of colliding tectonic plates. The Persian Gulf is in the foreground, with the narrow Strait of Hormuz at the upper left corner. Across the Gulf are the flat sands of the Arabian Peninsula. In the glint of the Sun, a portion of a river can be seen in the mountains, giving life to the nomads who raise sheep on these ridges. The swirls of an eddy can be seen just off shore in the Persian Gulf. South is up in this photo. (STS047-72-079)

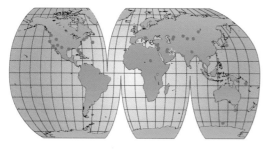

The Aral Sea

45 N 60 E

The border between Kazakhstan and Uzbekistan runs through the center of the Aral Sea. The area of the Aral Sea was the fourth largest of any lake in the world in 1960. Its area has since shrunk by 40%, exposing the gray areas of lake bottom on the left (east) of the photo and on the large island in the lake. In fact, the lake has split into two parts since 1985—a Space Shuttle photograph taken in that year shows the body of water at the lower left connected to the main lake. The Aral Sea has shrunk because the water from its feeding rivers, the Amu Darya and Syr Darya, has been diverted to irrigate cotton fields. Those along the Syr Darya are visible in the lower left portion of this picture as the dark cultivated land. The Amu Darya cotton fields can be seen at the top (south) end of the lake near the clouds. The Aral Sea is roughly 150 miles (250 km) wide. South is up in this picture. (STS047-103-043)

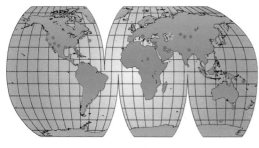

Amu Darya River Irrigation System, Uzbekistan

42.5 N 59 E

The Amu Darya River and some of the extensive irrigation channels that divert its water from the Aral Sea to the dark cotton fields of Uzbekistan can be seen in the glint of the Sun. In this case, irrigation was a good idea for a few generations, but now we are seeing the drying up of the Aral Sea and the dust storms that have resulted from this land use policy. South is up in this photo. Compare this photo with the image of irrigation in California's central valley. (STS047-88-098)

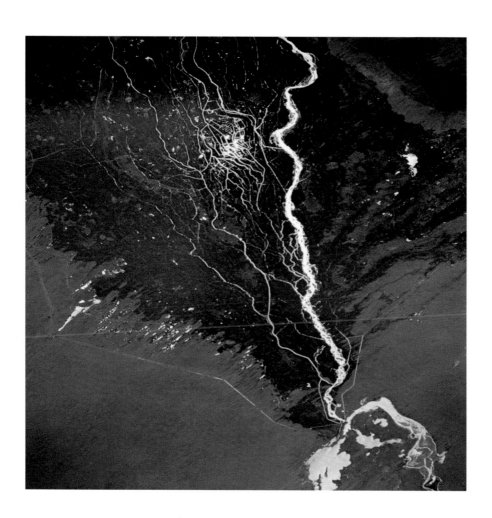

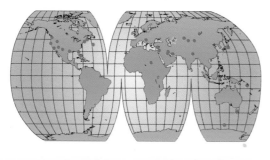

Aral Sea Dust Storm

45 N 60 E

When the winds blow over the Kazakhstan plains, they raise great clouds of dust from the exposed sea bottom of the shrinking Aral Sea. This dust is known to be carcinogenic. The sea salt, fertilizers from the cotton fields of Uzbekistan, and waste dumped into the Aral Sea over the years have come back to plague the residents of this area as the sea has been allowed to dry up, leaving only the poisonous dust. Some scientists estimate that the entire Aral Sea may dry up by the year 2020 if current practices continue. South is up in this picture. (STS047-87-070)

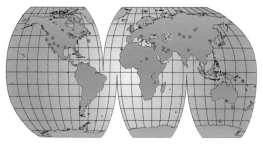

Lake Balkhash, Kazakhstan

46.5 N 76 E

This beautiful lake is in eastern Kazakhstan, roughly 100 miles (160 km) west of the Chinese border. It is 46 miles (75 km) across at its widest point, and 375 miles (600 km) long. Some of the light blue color may be due to the copper smelting on the lake's north shore. South is up in this photo. (STS047-89-009)

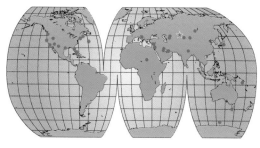

Lake Issyk-Kul in Kyrgyzstan

43 N 79.7 E

Even this remote lake in the Tien Shan mountains south of Alma-Ata and Lake Balkhash is touched by humans. The idyllic blue of the lake is marred by a lighter blue bloom of algae on the left (west) and bottom of the picture. The algae is a result of excess fertilizer runoff from the farms that dot the lake shore. The image is 104 miles (167 km) a side. (STS047-77-082)

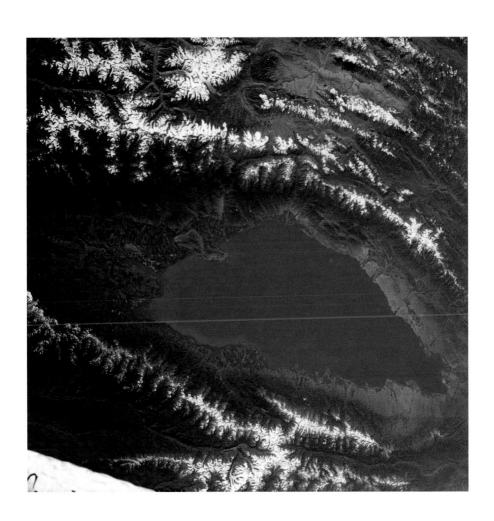

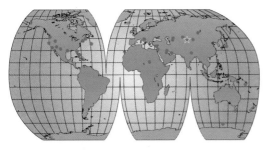

The Roof of the World

27.59 N 86.56 E

We took this photo as we were flying over central China, looking south over the Tibetan plateau at the highest mountain range on Earth, the magnificent Himalayas. The prominent valley in the foreground, the Rongpu Valley, points at Mt. Everest and its steep north face. Peaks visible in this photograph include Everest, also known as Sagarmatha to the Nepalis (29,028 feet), Lhotse (27,833 feet), Nuptse (25,784 feet), Baruntse (23,511 feet), and Bei Peak (24,879 feet).

The Himalayas (Sanskrit for "abode of snow") were shaped by a process called Plate Tectonics. The Earth's thin solid crust floats upon the mantle, a convective region of hot liquid rock, causing sections of the Earth's crust to move very slowly.

Two hundred and eighty million years ago, a supercontinent contained most of the solid crust that is light enough to float high above the heavier rock making up the ocean floor. It fragmented into large landmasses, called continents, that are still migrating the globe today. One such land fragment, India, began heading on a collision course for Eurasia (Europe and Asia).

As India approached Eurasia, it ploughed through the Tethys sea, a mid-latitude sea that separated the ancient continents, gradually squeezing the seabed and pushing it upward. Forty million years ago, India and Eurasia finally collided, thrusting parts of the landmasses and the Tethys seafloor five miles into the sky. Proof of this astounding process is evident in the fossils of extinct sea creatures found in some regions of the Himalayas. India continues to move north, raising the Himalayas at a rate of about two inches per year. However, this growth is offset by the same rate of erosion, so on average the mountain range is not getting higher. (S92-51560; enlargement of STS047-75-58)

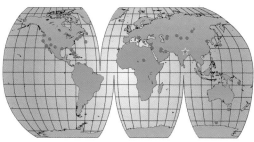

Lake Baikal, Siberia

52 N 104 E

Lake Baikal is the deepest lake in the world (in places as deep as a mile), and contains the most fresh water of any lake on Earth. It sits in a crack in the Earth caused by the collision of India with Eurasia, and is ringed by magnificent mountains. A huge cellulose factory at the south (right) end of the lake is causing air and water pollution in this remote area just north of the Russian–Mongolian border. The smoke plume from the plant is visible in this photo, as is the contrail of an aircraft flying over the Siberian city of Irkutsk. (STS047-51-01)

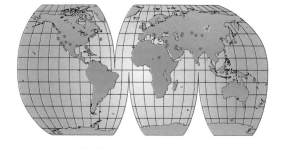

Turfan Depression, China

43.4 N 88.4 E

The cloud-covered Bogda Shan mountains in China at the top (north) of the picture reach heights of 18,000 feet (5445 m) and are constantly being eroded by rain, wind, and snow. Their debris cascades down huge fans into this area, the Turfan depression, which actually lies below sea level. The sediment from the mountains provides fertile ground for farms near the cracked and broken strata in the middle of the frame. These sedimentary rocks were originally laid down horizontally under water, but the forces that drive the movement of the Earth's crust have twisted them so we see them end-on, and have broken them in three places, where the rock was too brittle to bend any more. This area, north of the harsh Takla Makan desert, was on the northern silk road from China to Venice. This picture is 41 miles (67 km) a side. (STS047-91-054)

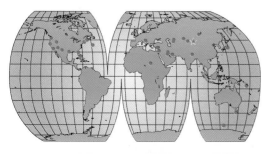

Tokyo Bay, Japan

35 N 140 E

The largest population concentration (27 million people) on Earth rings Tokyo Bay. The city of Tokyo sits at the northern end of the bay, and Yokohama (obscured by clouds) along the bay's western margin. The gray patches are close-packed urban areas; the brown valleys along the eastern shore of the bay are cultivated land, mostly rice. Although the Imperial Palace, at the northern edge of the frame, is cloud-covered, several other features are identifiable in the industrial corridor along the bay, including wharfs and Tokyo's Haneda Airport. (STS047-76-078)

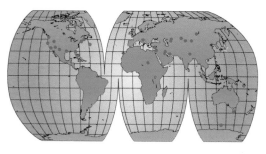

Aurora Australis

57 S 130 E

We now shift our view from the surface of the Earth to its beautiful atmosphere. Because we flew as far south as the tip of the Antarctic peninsula, we expected to have a fine view of the southern aurora. We did not expect to fly through it: to see it dance above us and below us, bright enough to light up the clouds far below, to close our eyes and have particles traveling at close to the speed of light reveal themselves with bright streaks of scintillation. We saw fountains, such as the red-topped one in the picture, appear for several seconds and then gracefully subside.

The green and red colors are due to the atoms of oxygen in the upper atmosphere: when they are struck by fast-moving particles, one of their electrons is knocked into a more energetic orbit or even separated from the rest of the atom. As the electron finally gets back into place, it emits a specific color of green (at a wavelength of 557.7 nm) or red (actually 3 colors: 630, 636.4 and 639.2 nm), characteristic of oxygen. The fast-moving particles come originally from the Sun, and have been trapped by Earth's magnetic field. They spiral tightly around the invisible lines of the magnetic field, and bounce rapidly from the North magnetic pole to the South. Where they enter the atmosphere, along a ring centered about the magnetic poles, we see the aurora. (STS047-20-15)

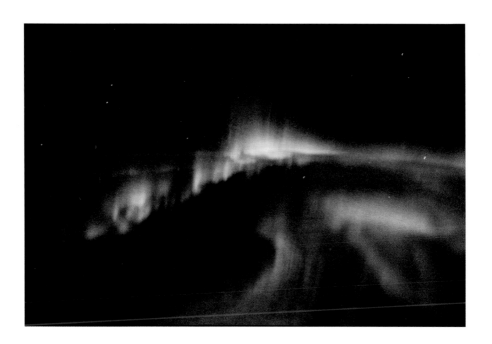

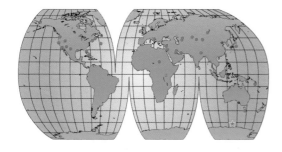

Aurora Australis and Dawn

45 S 147 E

Just as the Sun started to rise, we took a picture of the dawn (white glow to the left), the aurora (green curtain on the right), and the ghostly light from a region of the atmosphere called the "airglow layer" (reddish line parallel to the limb of the Earth). The airglow comes from atomic oxygen in a layer about 180 miles (300 km) above the Earth. (STS047-17-10)

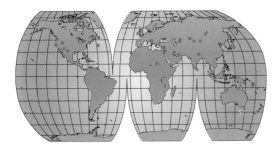

Sunrise with Volcanic Dust

15 S 159 E

Taken 740 miles (1250 km) northeast of Australia on September 18, 1992, this picture shows that the dust from the 1991 eruptions of Mt. Pinatubo (in the Philippines) and Mt. Hudson (in South America) is still in our atmosphere. The gray band in the stratosphere, above the thunderstorm, is the volcanic ash and dust. This dust has spread throughout the world, and has given us beautiful sunsets as seen from the ground. However, it has made the world seem dusty from orbit, and those of us who had flown prior to the eruptions noticed the difference immediately. (STS047-54-016)

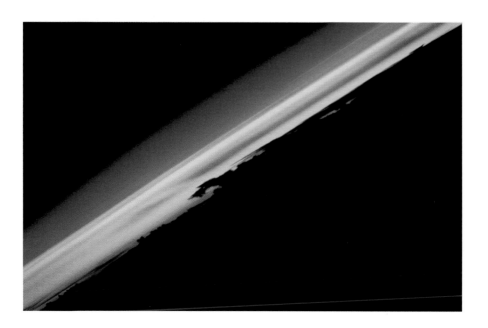

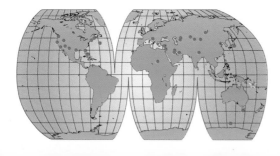

Sunrise

15 S 159 E

We had many opportunities to see sunrises and sunsets during our mission. This scene captures the silhouette of several mature thunderstorms with their anvil-shaped tops spreading out against the tropopause (the top of the lowest layer of the Earth's atmosphere) at sunrise. The colors of this picture provide insight into the relative density of atmosphere. The lowest layer (troposphere) is the most dense and refracts light at the red end of the visible spectrum, while the blues are separated in the least dense portion of the atmosphere (middle and upper atmosphere, or stratosphere and mesosphere). Several layers of blue can be seen; this stratification is probably caused by the scattering of light by particulates trapped in the stratosphere and mesosphere, particulates that generally originate from volcanic eruptions, such as that of Mt. Pinatubo, Mt. Hudson and, more recently, Mt. Spurr in Alaska. (STS047-54-018)

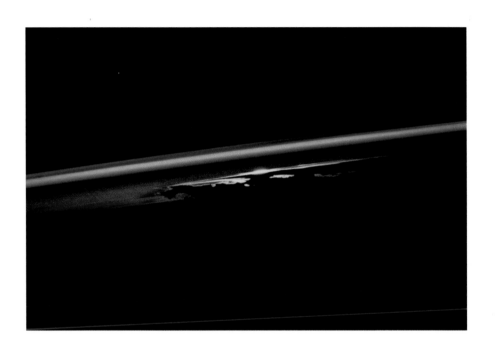

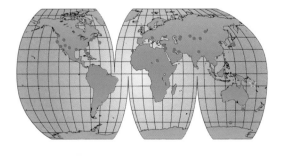

How to Order NASA Space Shuttle Photographs

Prints, slides, and transparencies of Space Shuttle Earth-looking photography are distributed through three agencies. A user may contact these agencies for ordering assistance, price lists, and order forms. To order a picture, you'll need to submit the Shuttle mission number, the film roll number, and the frame number, usually designated in hyphenated form, e.g., "STS047-90-101", where "STS047" is the mission number, "90" is the film roll number, and "101" is the frame number. These numbers are given for each picture in the book. The agencies are:

> EROS Data Center
> User Services Section
> Mundt Federal Building
> Sioux Falls, SD 57198
> Attn: Aljean Klaassen
> Phone: (605) 594-6151
> FAX: (605) 594-6589
> E-Mail: klaassen@dg2.cr.usgs.gov

and

> Earth Data Analysis Center
> 2500 Yale, SE, Suite 100
> University of New Mexico
> Albuquerque, NM 87131
> Attn: Amy Budge
> Phone: (505) 277-3622
> FAX: (505) 277-3614
> E-Mail: abudge@spock.unm.edu

and

> Media Services Branch
> Still Photography Library
> 2101 NASA Road One, Building 423
> NASA Lyndon B. Johnson Space Center
> Houston, Texas 77058
> Phone: (713) 483-4231

On-Line Catalog of Space Shuttle Photographs

Free public computer access to the NASA Space Shuttle Earth Observations Project (SSEOP) photography database of the Flight Sciences Support Office at the NASA Lyndon B. Johnson Space Center is available. This database can be searched to find photos of a particular area of interest, and returns a list of photo numbers, along with information such as the percentage of cloud cover, sun angle, lens used, and the time and date the photo was taken. Selected images can be downloaded in digitized form. This system can be accessed in four ways:

1. Via the World Wide Web:
   http://eol.jsc.nasa.gov/sseop.html
   http://ersaf.jsc.nasa.gov/sn5.html

2. Through INTERNET:
   Enter "TELNET SSEOP.JSC.NASA.GOV", and when queried for "Username" and "Password", type in "PHOTOS" at each prompt.

3. Through SPAN:
   Enter "SET HOST 9299", and when queried for "Username" and "Password", type in "PHOTOS" at each prompt.

4. Via modem:
   The modem can be 300, 1200, or 2400 baud; no parity; 8 data bits; and 1 stop bit. The number is (713) 483-2500. Keep in mind the following pointers:

—At the "CONNECT" and "CALL COMPLETE" messages, press the return key three (3) times in quick succession to get to the next prompt.

—When the screen reads: "connect 2400" you should enter the following: "SN_VAX". [Note—you must include the underline character between SN and VAX.]

—The screen will say: "calling 63111" and then "call complete" followed by a "#" prompt.

—At the "#" prompt, you should enter "J31X" and then press return.

Space Shuttle Photography Viewing Centers

The following centers receive microfilm and catalogs of all Space Shuttle photographs:

Alaska:
- Earth Science Information Center, U.S. Geological Survey
  4230 University Drive, Room 101
  Anchorage, AK 99508-4664
  (907) 786-7011

California:
- Earth Science Information Center, U.S. Geological Survey
  Building 3, Room 3128 (M.S. 532)
  345 Middlefield Road
  Menlo Park, CA 94025
  (415) 329-4390
- Map and Image Laboratory
  Davidson Library
  University of California
  Santa Barbara, CA 93106
  (805) 893-2779
- NASA Data Facility
  Building 240, Room 219 (M.S. 240-6)
  NASA Ames Research Center
  Moffett Field, CA 94035-1000
  (415) 604-6252

Colorado:
- Earth Science Information Center, U.S. Geological Survey
  Federal Center (M.S. 504)
  Box 25046
  Denver, Colorado 80225
  (303) 202-4200

District of Columbia:
- Library of Congress
  Geography and Map Division, Room B-01
  Madison Memorial Building
  1st and C Streets, SE
  Washington, DC 20540-4650
  (202) 707-6277

- Smithsonian Institution
  Air and Space Museum
  Archives Division, MRC 322, Room 3100
  6th and Independence Ave., SW
  Washington, DC 20560
  (202) 357-3133

Mississippi:

- Earth Science Information Center, U.S. Geological Survey
  Stennis Space Center
  Building 3101
  Stennis Space Center, MS 39529
  (601) 688-3544

Missouri:

- Earth Science Information Center, U.S. Geological Survey
  1400 Independence Road
  Rolla, MO 65401
  (573) 308-3500

Texas:

- Lunar and Planetary Institute
  Regional Planetary Image Facility
  3600 Bay Area Boulevard
  Houston, TX 77058-1113
  (713) 486-2182

Virginia:

- Earth Science Information Center, U.S. Geological Survey
  National Center Building, MS 507
  12201 Sunrise Valley Drive
  Reston, VA 22092
  (703) 648-6045